MY COVID-19 JOURNEY:
A Testament of God's Faithfulness

Iffy Maduewesi

DEDICATION

I dedicate this book to all Covid-19 survivors.

Yes! We made it. We beat Covid-19 hands down, ultimately.
To God be the glory.

ACKNOWLEDGEMENT

I hereby acknowledge the Omnipotence of God Almighty. I give Him all the glory forever.

I appreciate my husband, Emeka Maduewesi, Esq, and my children, Chiboy and Dezy Chia, for all their support, love and care, which led to my quick recovery. My gratitude also goes to my Cousin, Phillip Okeke for all his prayers and calls while I was sick.

On the same note, I appreciate my friends, Gene, Angie, Ify Obi, Dr. Loretta and Dr. Chichi, for taking over my domestic burden while I was busy combating Covid-19. Thanks, Oby, for the explanations on diabetes, the lungs and for the calls.

My special gratitude goes to my family and friends, who prayed fervently for my healing. I appreciate my Sis. in law, Dr Uzo Maduakor for making available those recuperating spices and my beloved brother, Innocent Okeke for delivering those spices to me here in the US.

Thank you Atty. Onyebuchi Okeke and my buddy, May-Chic, for reading through my book before publication. I appreciate you, Ifeanyi Okeke, for praying fervently for me and for those calls.

I also want to use this opportunity to thank all the doctors and nurses of Kaiser Permanente, Antioch, California, the emergency team of Kaiser, Walnut Creek, California, and the emergency team of Sutter Health, Antioch, California. Your dedication to your work inspired me to embark on this job. You all are amazing.

Thank you all for being part of my recovery story. Gratitude will forever be my attitude.

Iffy Maduewesi

TABLE OF CONTENTS

DEDICATION 2
ACKNOWLEDGEMENT 3
INTRODUCTION 7
   I HAVE A STRONG LOVE FOR LIFE, HUMANITY, AND GOD 7
CHAPTER 1: 9
TESTED POSITIVE FOR COVID -19 9
CHAPTER 2 24
MY HOSPITAL JOURNEY: The Tunnel Begins. 24
   My First Admission 24
   My Second Admission 28
CHAPTER 3 46
THE THIRD ADMISSION: Light At The End Of The Tunnel 46
CHAPTER 4 71
MY OUTLOOK ON LIFE: POST-COVID EXPERIENCES. OPINIONS AND APPRECIATION 71
CHAPTER 5 89
CONCLUSION 89
APPENDIX: 92
PICTORIAL GALLERY 92

INTRODUCTION

I HAVE A STRONG LOVE FOR LIFE, HUMANITY, AND GOD

Did I hear someone say that COVID-19 is a scam? If you still feel this way, please, have a rethink. This is a fight I fought, and I emerged triumphantly. COVID-19 is real. This virus now has a different meaning to me after my encounter with it. This singular occurrence in my life has changed my view about a whole lot of things; humanity, religion, race and even situations. This is a sickness I do not wish on anyone, not even an enemy. My sincere advice is that you change your views on COVID-19. This virus is monstrous. Do not wait till you encounter it before taking precautions.

I decided to give detailed attention to this write up because this obviously was the scariest near-death experience I have been through. It was a major occurrence in my life, and putting it down in writing is worth every minute I devoted to it. I had to write because this way, I share my experience and opinion on COVID-19 with the world.

Above all, I deemed this an appropriate media to proclaim the goodness of the Almighty God and make it indelible while I'm still here on earth. I want the future generations to glorify God for what He has done for me and the multiplicity of things He can do for us. I believe that God gave me a testimony to share, and I am being obedient to it. I certainly will not fail Him in this. He put songs of praise in my mouth, and I do not have a choice than to sing. I wish to declare His awesomeness. He gave me a second chance on earth and I am a happy camper.

CHAPTER 1:
TESTED POSITIVE FOR COVID -19

COVID-19 has never been inconsequential to me even without having a direct and personal encounter with it. It never was something I toyed with in any manner whatsoever. I was, and still am, one of those individuals who were fast to explain or advise anyone not to trivialize this virus. I made a couple of videos about the virus and instructions on how to limit the spread. I still have one running on Facebook to this day. I did a video for a magazine on YouTube on this issue. I was preoccupied with trying to create awareness and disseminate information to Africans back home. I had to, because at the time I did the videos, they were not taking it seriously. They called it all kinds of names; "White man's sickness, rich man's sickness", etc. So, there was a great need to create awareness and I did it in my own way. I called it 'contributing my own quota'. I did the best I could to assist family and friends at that time. I encouraged and urged everyone who could to assist others so people could stay home.

As an entrepreneur, COVID-19 impacted my business very adversely in 2020. There were incentives for businesses, but it didn't benefit me meaningfully because I had a couple of contract workers, not full-time employees. Like every other California business, I had to lock up for months. I was not so worried though because it was a general situation. As a person of faith and a believer in Christ, I'm so big on trusting everything in the hands of God, and this was exactly what I did. Apart from my business, my 25th wedding anniversary died a natural death despite the fact that all plans were already concluded. It was supposed to take place on March 28th, 2020, but was automatically canceled when California went on compulsory lock-down on March 17th, 2020. All these disappointments didn't hit me much because I wasn't an exception. It was all taken in good faith since it was nobody's fault. I despised the pandemic and it's cohorts and I pray fervently for it to be over soonest. This is one of the main reasons I decided to share my experiences so people who are still flirting with this virus should quit. It's time to get vaccinated if you've not done so because "a stitch in time saves nine."

Somehow, I felt so strongly that I had a part to play in stopping the spread of COVID-19. I so much wanted this to be over so I could return to my business. Mine is more of a summer business. I was scared that if I miss summer business, it will be winter again and that would not be good for my business. Thank goodness, retails started, and it was great all through till another temporary closure. Then came the riots, the fires, and the looting. All these happenings gave me the impression that this vicious virus wasn't coming to an end in a short while. I was so concerned about the violence and the length of the demonstrations because the virus could spread even more. This was another time I had to lay my feelings out there on social media.

I put forth great efforts enforcing COVID-19 guidelines. I still wear my mask. I have "Wear Your Face Covering" signs at strategic locations in my business. I placed hand sanitizers at strategic points too. I made soaps available at the restroom. I never let it run out before replacement like I usually do. I made washing my hands and reminding my kids to wash their hands a routine.

As a clothing retailer, I inculcated masks in my inventory. I sold all kinds of masks, from solids, prints to fashionable ones. I had disposable blue masks that I give out to customers who came into the shop without masks or to homeless people who just barged into the shop just to look. I have unapologetically spurned tens of customers from entering my business without face covering. Some stepped out to wear their masks and came back in while a few left. I have never been an advocate of something like I've been an advocate of "No Face Covering, No Entry." I see mask violation as being defiant because it takes nothing away from anyone. I consider refusal to cover up as disrespectful too. I understand forgetting one's mask and that's the reason I have free ones, but I don't understand being unperturbed about wearing a mask.

All these little efforts and precautionary measures I took in the past were the reasons I was all emotional when I tested positive to COVID-19. I was utterly dumbfounded when my diagnosis for COVID came out positive. I became very ill on March 13th, 2021. I was so sick and completely exhausted. All I did was leave home in the morning for my business and on getting to the shop, went straight to the office, crashed on double seats and turned on the individual heater so high. This was my routine for three consecutive days. The illness and fatigue persisted till we called the hospital to get tested for COVID-19. My whole family got tested, voila! We all tested positive.

We quarantined at home and took all the local steps we were told that could cure COVID, inhaling hot water with lots of spices and drinking lemon, ginger, garlic etc. We did all we could to wriggle out of this horrible situation. I could not really tell if this worked because things escalated without warning. I was so sick to even notice that I had lost appetite, sense of taste and smell. My teenage daughter, Dezy, woke up on the 15th of March, 2020 and came to my room and said, "Mom, I feel totally sick."

"What, mama?" I asked.

"I'm so sick and don't think I would be able to go to school," she replied.

I leaned closer and asked further, intently, "What's the problem, mama?"

She said, "This cough weakens my whole body."

"Okay then, no school," I responded immediately.

As at this time, we had not known our diagnosis. The COVID-19 test result came out on the 18th of March, 2021, and we were confirmed positive. Contrary to my character to always insist on school first, I did not encourage her to go to school. I quickly accepted because she would not play with such knowing me perfectly well. I called the school immediately to report my daughter's absence. I sent emails as well. The school was quick to recommend COVID test, but also concluded that whatever route we chose, she's due back in school on the 29th of March. That was an automatic quarantine order, with a request to update them with whatever happens. I am still surprised that I accepted my daughter's intention of not going to school without pushing back. I believe it was because I was already very sick and envisaged that she must have gotten sick from me since she was with me all that weekend.

That same day, my teenage son, Chiboy came back from PE bowling and indicated how he felt. He said, "Hey mom.
"I responded worryingly, "Hai Chiboy, are you okay?"
He answered, "No mom, I have been having this terrible headache all through my class."
I asked, "Have you eaten?"
"Not really. I feel tired and sleepy," he replied
So, I said, "Go rest a while, and we'll see what happens."
Feeling concerned, he said, "Okay mom, how are you, yourself?"
I responded, "I'm okay, I guess."

He went to bed and slept all day and woke in the evening with sound health. This was all the sickness he had because of COVID infection. Except for the loss of smell, my son's symptoms were the mildest in the family. His symptoms were not for long too, he recovered his losses almost immediately. All these symptoms manifested before we went for the test.

My husband was next in being seriously ill, but the differences were totally clear. He had chills, temperatures and sometimes, a little pain here and there but he didn't go through the whole nine yards. I saw his strength at this time as an act of God. I sincerely believe that God left him strong to help the family. He did all the runaround, making and answering calls. He made sure we ate. I really don't want to start imagining what would have been our fate if God hadn't kept one of us strong. Remember, our sickness is highly contagious, so friends assisting on a daily basis was completely ruled out. I am very convinced that God alone got this planned out just for His glory. Who would want to house COVID positive kids if we were both admitted in the hospital? There was a tribulation for sure, but God weathered it.

We were all pronounced COVID positive and had all the quarantine process read to us by the hospital and the County as well. My hubby answered calls hourly like a receptionist. Remember, it was four of us and each got his or her own share of calls, but he took all. More calls came from Kaiser Hospital since they are our healthcare providers. They wanted to know how they could assist. Kaiser sent us some sanitary and disinfectant items, and my doctor delivered a medication home to me that same day. We were told that there was nothing to do except be home and monitor our symptoms. We did exactly as directed. All these times, I was still exhausted with no strength. I tried to eat and keep hydrated.

God works in miraculous ways. We decided to contain this problem within us since none of our friends could come close or visit us because of the nature of our ailment. The moment we tested positive to COVID-19; we went on automatic quarantine. We concluded not to bother friends. As it was, I had the burden of sending a text message to a friend who lost a parent, so I sent out a condolence note to her. Somehow, I unintentionally let my feelings show a bit and a few minutes down the line, the phone rang, usually I was not picking calls because of fatigue but I had to pick this because I had just sent her a condolence message.

I managed to say; "Good evening, Gene."
"Good evening, dear. What's going on?" she asked.
"I just don't feel good," I responded.
Gene continued, "Your text message came in while we were having night devotion. I do not really understand the content. What's the problem? You scare me,"
I said, "Sorry, I just want you to know that I will get back to you on your loss when I get better."
She insisted, "I really do not understand."
"Well we all tested positive to covid-19. I can't really talk right now," I replied.
Immediately, she blurted out "God forbid, that's not your portion, Okay, later."

These were our first family friends to know about our predicament and it's a divine arrangement. Her husband was on the phone with mine at the same time. I had to drop almost immediately

because of shortness of breath. These couple delved into the matter instantly, got us our first set of vitamins and food as well. It is absurd indeed. How could we have been so oblivious not to think that we needed these vitamins at this time? This obviously eluded us, but thanks to our friends for being very thoughtful. I sincerely appreciate their kind gesture.

They sent us some fruits and food too. Any time I reminisce about this benevolence, I always wondered how we could have gone through this stage without this assistance and vitamins. I sincerely believe that God knows that we can use this help and created an avenue that invited our friends to the situation. Through them, our other close family friends got to find out and the rest became history. These friends made our problems their problems and took care of my family as I went through my dilemma. They made sure that my kids lacked nothing all through. These were friends indeed. I know I have a lot of valuable friends who would have done the same but had no knowledge of our situation because of this monstrous virus infection. We care for our friends and that was the reason for leaving them out of this ugly situation in the first place.

For your information, this was just two days from diagnosis. On this same day, I felt that I needed someone to talk to concerning my symptoms. Since I was still home, I needed some kind of advice. I remembered a good friend who is experienced in this, she's a nurse. I left her a message to call me at her convenience and she did. We discussed everything in detail. She went to the shop and bought a whole lot of things for me, ranging from thermometer, oximeter, to spirometer etc. She bought all manner of juices she felt we needed and sent them through my other friend while she was zooming off to work. My friend brought the items and joined the group that took food orders from my kids while I dealt with COVID-19 admission. There are so many kind people out there. God really did make a way for us.

On that same evening I got all this testing equipment, I tested my oxygen, and it was low, I left a message about it for my friend, Angie, and slept off. I was awakened by her phone call in the middle of the night. With sleepy eyes, I picked up the phone and said, "Hello."
Angie responded, "Hello sis. Please, go to emergency now. Sorry I was busy at work and didn't check my phone."
I said, "Aaah, now?"

She responded, "Please sis, go to emergency. You can call an ambulance."
Tiredly, I responded, "Where do I go?"
Angie said, "Go to any hospital close to you, John Muir, Sutter, Kaiser, any of them."
Worriedly, I responded, "Okay, thank you. Oh God!"

My husband was listening, but he wanted to know exactly what she said. He was asleep himself. He asked, "Iffy, what did she say?"
I replied, "She said we should go to the emergency room now."
"Right now? Which hospital?" He asked.
I snapped, "She said any of them. Let's go please!"
"Let me get the keys," my husband concluded.

We set off for the hospital. Kaiser, our health provider, was further away, so we ended up at Sutter Health. I still believe that starting with Sutter was the best move I embarked on. Meeting with this gentleman (nurse) that admitted me is one experience I will cherish forever. He reminded me of a phrase I have always known but this time, I felt it was a message for me. He said, "It will be okay. Just continue to cover yourself with the blood of Jesus."

I replied, "Thank you so much, this meant everything to me."

This message settled it for me. It sort of rekindled my hope and faith in Christ. It became my song all through my ordeal. .

Immediately they admitted me, I saw these technicians, nurses and doctors ran around tirelessly attending to me; blood work was done, oxygen, IVs, x-rays etc. All of this happened in a twinkle of an eye. I have never seen such coordination, my hope soared. I was very sick but was super impressed. I strongly believed that I would be fine considering the way they were handling my matter and because I have constantly covered myself with the blood of Jesus. No one could tell me otherwise. I was satisfied with the way things were moving. Then came the news of how my hospital, Kaiser, needed me to be transferred to their facility. They sent an ambulance over to get me. I was so scared to death. My hubby who brought me to Sutter had left already.

We had arrived at Sutter Health around 12am, and the Kaiser ambulance picked me up about 3.30am. The gentleman in the ambulance gave me words of encouragement to calm me down. He must have sensed that I was terribly agitated. We set off to Kaiser and we arrived safely. I was successfully delivered to the emergency team and the struggle continued.

The Kaiser crew proceeded from where Sutter stopped. The doctors and nurses were nice and compassionate. The doctor said to me; "You needed a CT Scan since you already had x-ray at Sutter Health." I waited patiently and I had a scan done as soon as there was space for me. The doctor said excitedly; "The good news is that there are no blood clots." This information was very consoling. The excitement in the doctor's face was very infectious. He was genuinely pleased with the result. It was great news to me too, though I didn't have a clue why the scan was conducted in the first instance. After the scan, they sent me to the 4th floor on admission. This became my first admission because of this deadly virus.

CHAPTER 2
MY HOSPITAL JOURNEY:
The Tunnel Begins.

My First Admission

The nurses and doctors I met at Kaiser were so kind and very nice to me. They were doing all they could to make sure I lived. I was busy inviting God while trying to keep awake. For some reason I could not explain, I was scared to close my eyes because I did not want to die. I got drips and blood work. All manner of medications coming through. They all made me feel that I will be fine in a short while.

In the morning, I was astonished when they referred to me as diabetic. I could not believe my ears because I have never been diabetic all my life. This became the first major problem. The doctor pleaded with me. I can still see her face when she said, "Please, accept it. You were diabetic before the COVID infection."

Vehemently, I opposed the statement, saying "I won't accept this, doctor."

"Please, don't be in denial," the doctor maintained.

I responded, "I'm not in denial, doctor. How could I be a diabetic without symptoms. I am now diabetic because of steroids medications; I know I will get back to normal after my treatment." Doctor responded, "We will check your A1C and it will confirm this."

The doctor and I argued on this almost on daily basis but I refused to accept it. It was an amiable dialogue, though the subject of discourse was intense. We argued with maturity and respect. She just didn't want me to believe I became a diabetic from their treatment. Personally, I honestly did not care what they used to treat me, all I cared for was to stay alive first. Checking my blood sugar on a regular basis and getting insulin became part of the routine. I found out somehow that some medication used for COVID does spike blood sugar. That made my case understandable to me but the doctor didn't think so. I became so paranoid for this sole reason and battled this throughout my sick time. How could I have been diabetic with no symptoms? This never added up to me.

Though I hated the word diabetes, I was very obedient to all the instructions. I did my breathing exercise, walked around the room too. There were no visitors, so I was minding my business. The nurses were so nice, they were always in the room, so loneliness was out of the equation. I drank hot water all through my stay at the hospital and I made sure the nurses got that for me. I felt trapped down in that dead zone and all I did was stared at the ceiling and trusted God. I was so confused with this sickness that I obviously had no idea on what to think. I read all the literature on the wall countless times. I really lost my sense of reasoning in my first admission. My psychology at that time was in total disarray.

I was at Kaiser for three consecutive days and the doctor told me about my discharge because I was done with my IVs, my oxygen level had become better, and I was stronger. The team that worked with me really believed in me for some reasons. They felt that I did great in every way. When the nurse pointed at the spirometer to remind me, I simply responded. "I suck at this exercise."

The Nurse replied, "No, you are really good at it."

I responded, "I don't really think so, when I can't get to ordinary 1500."

The Nurse replied, "You actually did better than most patients here. Some can't even lift that up." "Really? That's encouraging." I said.

"So, don't bother yourself. You're doing a great job," the nurse replied.

I really felt that I wasn't doing good with the spirometer exercise, but the nurse urged me not to worry but to be steadfast at it. I was really encouraged after this conversation and I put forth more effort. The doctor felt that I was good enough to be discharged. My daughter and husband were invited to learn the 'new normal' of diabetes. After the teaching and learning section, they packed us loads of medications and sent us home. I was so happy I left the hospital alive. My daughter became the home diabetes nurse.

My Second Admission

I got home and followed my prescription guidelines strictly. I adhered to all the advice that was given to me. I made conscious efforts not to violate any precepts, so I do not go back to the hospital. I ate right too, little portions of balanced diets. I ate fruit snacks in between. I thought I got it all covered but little did I know that it was not over. We even had a friend drop in briefly to pray with us. It was in the process that the tragic incident occurred.

It was the third day after I left the hospital, I had this excruciating pain from my lower back to the front of my chest. I screamed; I thought my skeleton would fall apart in bits. It felt like I was having cardiac arrest. We checked my oxygen and it fluctuated so widely and we knew it was an emergency situation. So, we set off to Kaiser Hospital. We registered and they took me to a room. This was where I saw them worked so tirelessly like the first day I went to emergency. A lot was done in a heartbeat. I had an X-ray and CT Scan within a short interval. I was given an anti-inflammatory medication that stopped my pain instantly. The nurse who delivered the medication had said, "Everything will be alright. We'll take care of you."

Everything was moving like lightning until about 30 minutes after the Scan. It was like a decision had been reached about my case. Even this caring nurse that told me that everything will be okay soon became distanced, but I could still see him in the hallway. For what felt like eternity, no one came to me. I used my favorite phrase several times, but I refused to close my eyes. I waited for the worst. At last, a doctor came in and talked as calmly as he could, he called my name as he dropped the bombshell, "According to the scan, you have blood clots in your chest and there's nothing anyone can do about it." I took a deep breath, I did not understand what a blood clot is doing in my chest, but I know it means there's imminent danger. The doctor went on, "You need to have an ultrasound in the morning, so we find out if you have the clots in your legs too."

I listened without uttering a word for lack of what to say. Since nothing can be done here, I

surrendered it to God in prayers. We concluded that there will be an ultrasound in the morning. "I will not have clots in my legs," I muttered to myself. He left and that nurse who had worked so hard all evening came back. He managed to say, "You already heard you have blood clots." All I said was "Yeah." He only stayed away because it wasn't his duty to break the bad news to me. After a long time and no one came to me, I started waving for attention. When someone came closer, I asked what was next for me and they told me that they were waiting for the admitting doctor. That is the longest wait I've been through in recent times. It seemed infinite because of my disposition. Then, another doctor approached me with the weirdest questions anyone has ever asked me. First, he asked, "Who will make decisions for you if you couldn't?"
I responded, "My husband."
Then he asked a further frightening question, "Do you want us to try to resuscitate you in case you pass out and have no breath?"
Worriedly, I responded "Yes."

I felt I had no option. I wanted to stay alive. These questions at this inopportune moment were the worst questions I have heard in my life journey. To me, it was a confirmation I needed to prove my worst fears. I already felt that something was awry. I felt somewhat that something like death was seriously knocking at the door, but I kept covering myself with the blood of Jesus. I felt so miserable at the end of the session. It wasn't long before the admit-doctors came, and they took me to the fourth floor and gave me a room. That was a welcome change.

This became the beginning of a new dawn for me. I sincerely believed that subsequent events were divinely orchestrated. They handed me to a nurse whose name still rings a bell in my head. To be precise, I called her an "angel" because she has all the attributes of one, but her name is Allison. I was already so battered with the news of the blood clots. I was so broken and needed a shoulder. This angel of a nurse, gave me a shoulder, mended my wounded soul without me demanding for it. She started by asking my name and introducing herself, getting me all I needed to get myself cleaned out. She offered me food, got hot water for my flask. She took care of me like she knew what I was passing through at that moment. Of course, she knew that I was a COVID -19 patient, but I meant my emotional and psychological state at that minute. She made me so comfortable, and I was able to ask her questions concerning blood clots and she explained it in a very simple way, and I was able to understand it. It's absurd now, but then, I asked her, "Nurse, please, am I about to die?" Softly, she responded, "No, not at all."

Again, I said to the nurse, "When everyone stayed away, I perceived they were scared to tell me the obvious."

She replied, "No, the doctor will give you blood thinner to manage the clots."

Quickly, I responded, "So, this won't kill me?"

She replied, "You will be fine, don't worry. You need to sleep and rest for now."

Feeling a lot better, I said, "Thank you so much, you are God sent."

Before she left, she said, "You should always lay tummy-down so you can breathe better to help your lungs to heal."

I was so happy and became hopeful again. She helped me to lay the right way on the bed. When she saw that I was calm and she was

done with her work in the room, she gave me the phone, dimmed the light and left an instruction;
"Here's the phone, there's my number on the wall, call me for any reason whatsoever."
I nodded in response, as I was already so exhausted.

I could not believe how kindness could change things instantly. I was so at peace with myself that I felt that God deliberately sent me some bliss amidst my wildest storms. I still believe that to this day. Guess what I did after this reassurance from this spectacular nurse? I calmed down and slept like a baby. Just to refresh your memory, I only sleep when I have no other option for the fear of closing my eyes and passing out. But this time around, I decided to close my eyes. I did, and slept deep with much bliss for someone in my situation.

I woke up the next morning, strong and happy. She was by my bedside but this time to say goodbye because her shift had ended, and she was set to head home. She promised to pray for me at home. Before she left, she introduced the next nurse on duty. The new nurse started off admirably with me until she came up with the woes of blood clots. She erased all the great memories Allison had imprinted in me. The new nurse said, "I have a video on blood clots for you to watch."

I stuttered, "Yea, what for?"

I went further to ask "Why do they want to check if I have clots on my leg?"

She replied, "They need to know because the clots on the legs are usually very dangerous."

I said, "Really?"

"Yes, because blood clots on the legs travel and they can cause cardiac arrest and other problems," she replied.

I responded again, "Really? I'm so scared."

The nurse said, "You should not be scared, the doctor will prescribe some blood thinners for you. You have to shoot it and it will take care of the blood clots."

I screamed! "Shoot, like insulin?"

"Yes," she affirmed. "You have to constantly check your blood to know what unit to shoot."

I was like "Wow! God help me."

The Nurse then said, "Here's the video, you will learn much from it."

I became very despondent. I could not believe my ear. I was already injecting insulin for blood sugar, and now, blood thinners? I was consumed in my thoughts and asked lots of rhetorical questions in my head. "Will I be injecting blood thinners on myself? How can I be going to the lab constantly to check my blood every two days? The truth is I would prefer death than going through these two. What is life then?" I became very gloomy and really began to weigh my options.

At my state, I felt that playing the blood clots video for me to watch and then leave, was totally unacceptable. She left a few minutes into the video and did not return till after the end of the video. I can still remember asking her if I will ever recover from the blood clots and she said that the video was more for people who had surgeries and are healing but not COVID patients. She said she doesn't know their fate. I went through this agonizing video all alone. I bet she believed that she kept me busy with it. All the hopes I basked in the night before were all dashed. I am sure that she must have answered innocently but the answers were like daggers to my heart. I stayed there worried about my life with no one to talk to but God.

At this point, I remembered my kids and cried. I cried because I am human and not because I lost faith in God. I mostly thought about my kids and what they will go through in life in my absence. I was totally down and waited impatiently for the doctor who entered my room within that period. The doctor spent more time than usual in my room explaining my circumstances. I was in despair already. After exchanging greetings, I asked the doctor, "Doctor, what's next for me?"
Doctor responded, "Remember, you have an ultrasound for your legs scheduled for this morning."
I replied, "It's about 10am now but I've not heard anything."
The doctor said, "Once there's space for you, they will notify you"
"Okay doctor." I responded, and quickly asked, "Is it true that I will be checking my blood regularly and shooting blood thinners on myself?"

The doctor replied, "No, where did you learn that from? I will give you a good tablet and you will not go to the lab for one day."

I asked again, "Doctor, will I take blood thinners for the rest of my life?"

He responded, "I give you two months, three months at most, to recover completely and you will not be needing any medication."

Feeling unsatisfied still, I pushed further, "Doctor, please be honest with me, I can handle it."

He replied, "I answered you honestly."

Feeling relieved, I heartily said, "Thank you so much for your time, doctor."

The doctor told me that they were still waiting for a space for me. So, I waited patiently.

The doctor rekindled all the hope Allison initiated the previous night. He bulldozed all my fears and brought me back to a positive reality. I became optimistic again. There were other people with him, but he still spent a lot of time explaining these things to me. I am certain that he saw the pains in my eyes. This was one of the most agonizing periods in my admissions, but it changed after our dialogue. Immediately after the questioning and answering session with the doctor, I came out of that bad mood. I was very satisfied with his answers because I pleaded for sincerity and he told me the truth and I became hopeful again. I was so ecstatic with the compassion I got from him. There was no doubt that he would want me to experience the quickest recovery. His extra verbal expressions as we discussed communicated that to me. There are good doctors who mentally step into the shoes of their patients and feel their pains. This doctor is one of those.

As soon as the doctor was exiting my room, the nurse who had me watching the blood clot video came in and started applauding the doctor on how he humbly responded to my questions and the time he spent convincing me that everything will be alright. According to her, she was so elated because of how he patiently explained things to me, and so was I. I noticed she was watching from the door.
I asked her, "Why did you have to play the depressing video and leave for a long time?"
She replied, "So sorry about that, I was with a patient.
Understandingly, I responded, "No problems, I get that."

She was gone for that long knowing how sad that video could be but since she was busy helping another patient, I sincerely understood. There were other patients too. Then I reprised the doctor's explanations concerning blood clots and blood thinners to her. I was quick to point out that it would not be an injection, I do not have to go for a blood test even for once and finally, it can be cured in a matter of a short time. I had to explain all that to her because I sincerely did not know how and from where she concocted her own story. Though the doctor said that those were old practices, she needed to be up to date. I sure narrated all these to her, just like the doctor explained to me.

She was my nurse for the most part of the day. She's considerably a great nurse despite the blood clots issue. She was smart, very caring and did every other thing excellently. I was invited for my leg ultrasound when there was a space for me. I was wheeled down, had the ultrasound, and was wheeled back up to my room. It was very painful, that thick stick against my leg bones gave me excruciating pain, but I already sacrificed my body to pains, but not my soul. I had uncountable pokes for IV, blood drawn, shots etc. I taught my body to adjust and simmer down for pains. Like I was saying, I went back up in a wheelchair. In about 30 minutes, the compassionate doctor came back to tell me that I have no clots in my legs. He went further to console me and said that the position of the blood clots in my heart was good enough. I cannot precisely lay my hands on how he said it exactly, but I can ascertain that he meant that there is something beautiful about my ugly situation. The news was a very big relieve. He encouraged me and I was hopeful. I am sure that all he wanted was to console me and he did a good job actually. He told me that I will be discharged the next day and I was a happy camper again.

The same nurse attended to my every need till her shift was over. She introduced the next nurse before she left. The new nurse did excellently too. I received good care and attention throughout the day and night and was recovering rapidly. I had other nurses come in close succession for one reason or the other. To draw blood for labs, for blood sugar checks, for respiratory therapy and other reasons. Lest

I forgot, I used inhalers too. I actually started using inhalers from my first admission, but it became very noticeable to me this time around because of the frequency. It had never occurred to me that one can have up to six puffs at a time, but I learnt from personal experience. I have never had even a single puff of inhalers in my lifetime but here I was, doing those extreme numbers. COVID infection really had a combat with me but I emerged a winner to God's glory. I am proudly a COVID-19 survivor. You cannot tell me otherwise.

It is that time again, my discharge time and I could not wait. My husband took a minute to get to the facility since the decision to discharge me was sort of impromptu and was relayed to him a little late. He had other engagements at home with the kids but had to rush things. He finally arrived at the hospital, went to the busy pharmacy, and got my prescriptions.

At this time, I had this nurse with a great sense of humor to discharge me. He took off the IV on my hands and screamed at the huge needle that looked like a hammer in my hand.

He screamed so loudly, "What is this? I've never seen a needle this size."

It took him unawares, he continued screaming, "How did they come up with this?"

I responded, "I guess it's because they needed it for a scan. They used ultrasound to find my veins."

The nurse replied: "Okay, let us get you ready to go."

That needle really scared me too. He said that he had never seen such a big-sized needle and that was the same with me, I have never seen such till that day. What happened was that the night I was admitted, the nurse had difficulty finding my vein for IV and a technician was invited to use the ultrasound to find my vein. Since they needed that for the CT Scan too, the technician decided to use the biggest needle ever for that IV. I did not notice till my discharge process. The nurse could not stop screaming, though. If I had seen it before it was used in my hand, I wouldn't have allowed that. I cannot believe that someone put that in my arm.

Well, it was time to go, and he got help to wheel me down. My hubby was waiting already, I joined him, and we drove home. I praised God and was happy to still belong in the land of the living. They all bade us goodbye and asked me not to come back for admission again.

Home, sweet home. My kids were waiting for me. They were so excited I made it home the second time.
My daughter was like, "Welcome home mom, I missed you. I heard the garage door open and I came downstairs."
I responded, sounding very tired but excited, "I'm fine mama, I'm so glad to be home."
I washed my hands in the sink and touched her face and asked;
"How are you mama, you're okay?"
She responded, "I'm fine Mom, I was worried about you, I'm happy you made it home."
As we were talking, my son, Chiboy, joined the conversation.
He hugged me tightly, and said, "Welcome home Mom, I really missed you. Tell me all about you?"
I responded, "I am alright now to the glory of God. I was worried about you guys actually."
He went on to say, "Dad told us about the blood clots."

I replied, "Yea. But I'm alright now and I'm home."

We went on and on, but it was a happy reunion. I felt blessed and loved at the same time because my friends made sure that there was lots of food in the house. My family didn't lack anything in my absence. I continued my medication. I got an insulin and blood sugar check the first time I was admitted, now I added blood thinner and pain medication to the lot. My daughter administered the blood sugar department while my husband handled the pharmacy department and every other department. They really took good care of me, more than I expected. I honestly did not miss the excellent care the nurses gave me. My spouse was very patient with my situation though I am an amiable patient, I must say. I tried to make it easier for them. They did great. My son was the consultant, always asking how I was doing. He was always patting my back while hugging me with a big, "I love you, Mom." Gratitude will forever be my attitude.

The blood sugar check was brand new to us and we despised it but had to adapt. The constant subject of discourse in my household at that time was how to go back to my former self before the COVID infection. My daughter poking me four to seven times a day just to have a glimpse of blood is not pleasant at all, but we continued doing it till I got back to my pre-COVID self. I did not have an appetite but had to eat for all the medications. It felt good being at home. I tried doing all I was asked to do, breathing exercise, checking my oxygen, walking a little bit, etc. Nothing much to do, the TV News was very depressing, so I stayed off the television. I had no strength to focus on the movies I love either. Frankly speaking, I was contented to be home with my family. And everything seemed to be moving alright.

CHAPTER 3

THE THIRD ADMISSION: Light At The End Of The Tunnel

I was indeed thankful that the hospital kept calling and checking on me while I was at home. It was at one of those hospital checks to find out about my oxygen that the doctor noticed that I was having a short breath and we decided to check my oxygen. I did and the numbers were not good, and she asked me to go to emergency immediately and this became my 3rd admission. I hated these emergency visits, but I had to go. My hubby and I set off to the hospital. This time around we decided to go to the Kaiser hospital closest to us since they always say you should go to the closest hospital.

We arrived at Kaiser, Walnut Creek, CA, around 11am. I stood in line and waited for my turn. I thought I would collapse at that line. I did not want any attention because everyone that was on that line had an emergency reason that prompted their visit. Finally, thank goodness, it was my turn. The sorry but funny thing about my illness is that once I pronounced my COVID status, they pointed to a lonely, isolated corner for me to go and that's exactly where I went. We went over to the door and waited. I was freezing till my husband had to take off his jacket and hand it over to me. That jacket meant the world to me at that instant. Remember, in all of this "I'm covered with the blood of Jesus" didn't elude my lips. We waited for a while before someone came in and told my hubby that he was not allowed in. I handed him back his jacket and went inside.

Then the lady asked me the reason for my visit, and I told her. It did not take long before this gentle doctor came and sat down. He spoke to me with so much care and love. He immediately addressed my circumstances and I felt totally lifted that I thanked the doctor for uplifting my spirit. He demonstrated his acceptance by putting his hand on his chest and bowed and responded that my words warmed his heart. I was looking for encouragement and he did justice to it. The doctor assured me that they will find a room for me and deal with my matter. In a few minutes, they got me in a room and descended on my case in earnest. It was like a fight.

Everything was moving so fast. When they saw that my whole body was so bruised from past admissions; lots of IV openings, blood drawn etc., they applied more caution. They became extra careful with the way they handled me. I heard one nurse say that it was hard to find my veins, but she is not poking except, she is sure. They were careful with me. They did a lot in a short time; from drawing blood, drips to x-ray, and more. Then, the nurse told me that she needed to get my nose swapped for a COVID test. She said they had to do it because I have been on several admissions. I accepted but it hurt so much. She got the swap and went away. I waited in the room till I noticed a change in the atmosphere, and I was certain that things were not right. As I was contemplating on this, a nurse spoke so loudly, "You tested positive with another strand of COVID." For sure there is a problem, my thoughts were right. Everyone was calm, the tempo was so different. I did not bother to ask for clarification. I really felt for the nurses because they were gladly attending to me without their gear knowing that my quarantine days were over, without the slightest idea of what they were up against. Everyone stayed away for split seconds, but they immediately geared up and came to my aid.

They came for more blood. They had me on IV and were busy with other things. It was at this juncture that the phone rang, and I had to pick it up. The doctor that called said she was sending an ambulance to pick me up and to transfer me to the Kaiser Antioch facility where I observed my first and second admissions. She insisted because, according to her, they had limited admission space at that very facility. I was super enraged because I was at the Kaiser facility closest to my house, and I explained that to her. I even told the doctor that called while I was home that I was going to this facility and there was no objection. I was sad about this. Going on admission in the same hospital for the 3rd time? I had no option here, I thought. Meanwhile, the nurses were busy administering medications on me. They attended to me like they had no idea that I was leaving that facility. I saw people who were busy working for me to be alive, I kept quiet and enjoyed the hospitality and was hopeful. At that moment, my husband had already gone home, so, I had to deal with the transfer of facility business all alone.

The climax of this emergency visit was when the ambulance crew arrived to pick me up for the transfer. The nurses blatantly refused to hand me over to them. They asked me if I had knowledge of the transfer, with much concern though. Of Course, I was called and told but it sounded like I had no choice here. The doctor that called didn't ask for my opinion. The nurses were forces to reckon with. They refused to hand me over to the ambulance crew. They vociferously told them that I had IV on and will not go anywhere unless they have a licensed nurse riding with us. They were left with no other option than returning back in two hours and that was exactly what they did. It appeared to me like they did not see the reason for the transfer. I was calm and fine because I was getting my treatment while all these went on. As bad as it may sound, the ambulance crew picked me up two hours down the line after my IV were all gone. Then, we set off to Kaiser, Sand Creek, Antioch, CA, facility and arrived safely. The journey felt longer than I anticipated. It felt like eternity.

We waited outside a little bit after we arrived at the premise. The only consolation was the fact that I did not have to go through the emergency or admission process again but was taken straight to my room. Though I was fuming because I had to travel all those ways to this facility, it felt like home a little bit. I have been admitted here twice and being confined on the same COVID floor; I had seen most of the nurses before. It was so unfortunate I had to come back. They asked me to recover so I do not have to come back on admission, but here I am again. I came with nothing, but I held onto the blood of the lamb. I heaved a sigh of relief when I entered my room. I requested permission to use the restroom before they put all those gadgets on me. They already plugged the oxygen, but I asked to be excused. Almost immediately, the doctor that started this transfer shenanigan called and was glad we arrived safely. I tabled a few complaints and received nice responses.

The truth was that I got excellent services from Kaiser, Antioch, CA, both in my first and second admissions. But this last admission has no comparison. I saw nurses and doctors working extremely hard as a team so that I could live. The doctor told me straight on that I tested positive to COVID-19 a second time and now, I added pneumonia to my sicknesses. Remember, it was first diabetes from steroids, later, blood clots, and now, pneumonia. They told me that viruses do not need antibiotics for treatment, but pneumonia does, and I had to take some through the IV. They said that I had a temperature, and my blood pressure was a little elevated. These were not the case in my first two admissions. This very admission was obviously more serious than the first two.

At this point, I have already gotten to my wits end. I could not take it anymore. I broke down in tears and for a second the thought of dying and leaving my kids behind to deal with the world by themselves crept in again. I was overtly terrified. It was like fiction, but clearly translating to reality. Amidst my disillusionment was God. Even at this point I still trusted God. I believe in Psalm 41. I see myself as a beneficiary of the Psalmist's prayers in Psalm 41 because that is who I am or who I aspire to be. I sincerely believe that I have a special place with God. I know that God has been mindful of me all my life. I know lots of people smile because I exist. I strongly believed that God will not let me die except there is a force bigger than Him, and since there's no such force, I knew I will live. I only cried because I am human.

I resumed the first night of my third admission with this kind-hearted nurse. We only spent a couple of hours and her shift was over. She introduced the graveyard nurse to me before she left. This graveyard nurse was phenomenal, I must say. Immediately she arrived, she took over for real. She straightened out everything, changed my dress, my socks, and my beddings. Got a moisturizer and massaged my feet. She asked me to brush and use the rinse. She asked if I wanted anything, but the other nurse had already given me food and all. This nurse treated me more like family. She taught me how to lay face down so I could breathe better and help my lungs. Each time I complained to her about my pains, she would change my position. She said it's psychological. It got to a point, I insisted that I needed pain medication. She called the doctor and got one for me.

She took care of me just like Allison, the previous nurse who took care of me at my blood clots admission. I have tried to remember her name, but couldn't. That is unfair to her because she was my graveyard nurse for three straight nights, and we bonded so well. I looked forward to her shift. I guess I was not familiar with her name before I met her. This nurse gave me the most remarkable attention. She showed she was always happy to see me, and that was the same with me. Once she came to work, even before she put her belongings away, she would stand by the door and ask me.
"What do you want me to get for you once I change, Ifeoma…What do you need?"
She asked so many questions at a time and my only response was,
"Please, get me hot water and Chamomile tea. She's like, "What supplies do you need?" I replied, "I still have everything, thanks so much."

This nurse was unbelievably amazing, and she repeated this every night. I sincerely appreciate her kindness. I felt loved. I started looking forward to her shift. Since I spent four nights at the hospital, I got to spend three straight nights with her. She made me feel very comfortable. I was
very comfortable with all the nurses though. This was God's favor at work. His grace was sufficient for me. If not God, tell me how I managed to attract all the love and care from everyone?

Still on my third admission. It was like I just started because of a new diagnosis; pneumonia. There were lots of things being treated at this time and all I did was stick to instructions. There was this doctor that knows how to make sure that I keep to instructions. She came around to see how I was doing. My room was busy because of so many routines that transpired daily, different nurses for different reasons. Other than the nurses that were assigned to me, there were other nurses for a lot of other things. The division of labor was remarkable.

I spent more days at this visit than the other admissions. I had to take most of the antibiotics through IV. These doctors and nurses worked so tirelessly that I believed they knew that I would recover. I could feel the energy in the air. Most of my medications doubled at this visit. The inhalers that I used to do two puffs, four times a day became 4 puffs, four times a day. I have never had any reason in my life to do inhalers, not even for a common cough. Same with every other medication. Even the insulin added 15 units of slow release, but I declined that. I seriously frowned at the extra insulin because I feared that with a lot of insulin in my system, I will become so dependent on it and will not be able to get myself to my pre COVID self. So, I told the doctors that I was not going to do it and I did not. One day, I decided to Google steroids induced diabetes and read all I could find there. I then found out that these doctors were only trying to make sure that I would not have any complications. They were being extra careful. I became a little remorseful. I am indeed so sorry for being a little headstrong on this. I didn't regret not doing the 15 units and everything worked together for good.

On the third day, the doctor came and spoke to me pertaining to my medications and.my discharge plans.
"If all things become equal, you will be discharged tomorrow." He said.
I was excited but softly replied, "I'm really not in a hurry to go home, I'm more interested in getting completely healed so I wouldn't return here on admission"
The doctor quickly replied that my white blood count is looking good. From then on, it's all about my discharge. Every plan was geared towards my discharge. There was a long deliberation about Kaiser giving me oxygen equipment to take home, but we later resolved that quagmire. We arrived at a better concession to that effect.

Then came the other doctor who delivered bad news before. Remember the doctor that asked if I would like CPR if I were out of breath? Now, he is insisting that I was diabetic before I had COVID. This is a story for another day. This time around he came with another question pertaining to what will happen if I die. The only problem was that I had no mental record about the questions this period. If I knew that I would share my experience, I could have written certain things down. In summary, all his questions told me that I had a thin line between life and death. If I had died in the process, it would not be a surprise. Trust me, I loathed seeing that doctor. I understood he was doing his job, but I would have preferred if he never came around. Great, I was not asked to review him because I would have been honest about it. Well, someone has to do the ugly part, I suppose.

I had a nice night under the care of this amazing and overly protective nurse. I just must give this up to Kaiser. It was great ordering from the dietician. They did a spectacular job too. It was delightful to hear the phone ring, and when I picked it, the caller went on.
"Hello, I'm the dietician at Kaiser, ready to take your food orders."
I replied, "Hello, I don't understand."
The caller proceeded, 'Say what you want to eat
for breakfast, lunch and dinner."
Naively, I asked, "Things like what?"

He replied, "For breakfast, cream of wheat, toast, fruits, eggs…. "For lunch chicken, soup, noodles, vegetables, fruits..."
I responded, "Oh, I love cream of wheat."
He inquired again, "What for dinner, salmon, chicken breast …."
I politely interrupted, "I want salmon and noodles soup. Please, could you choose for me?"

Every time, someone would call daily to ask me to order all my daily food. I usually didn't know what to order because I had no appetite and couldn't really taste things, but I still ordered stuff. In my mind, I was like; "what kind of restaurant is in this hospital?" They even asked if I wanted salmon, chicken, etc. I always said salmon because I needed something that would dissolve easily even if I did not chew. Otherwise I would simply say No, but some of my medications require that I eat some food. One faithful day, they asked if I wanted salad and I ordered chicken salad. When they brought the food, it was like three persons portions There must have been a miscommunication somewhere. How could they even expect someone who's sick to eat all those.

Now, it was my last day at the hospital. The clock was ticking, and I am sure that I was so exhausted. I became highly irritated. My frustrations were obvious to anyone that attended to me on this last day. The doctor added bowel movement as part of what we were waiting on before my discharge. My problem continued but worsened when the nurse came with this excessively sweet liquid and asked that I drink it. Without mincing words, I told the nurse that I did not need it. She insisted and I spurned the plea. She left the bottle on the table and walked out of the room frustrated. I knew I had no option than to take the liquid if I plan to leave the hospital. So, I gulped that drink and fell asleep instantly. I could not fathom the reason I fell asleep, but I slept so deeply and woke up very upset. I asked the nurse what they gave me. Momentarily, I felt like someone in a psychiatric home where they use sedative injections to manage patients who act up.

As I was waking up, the nurse entered the room and demanded that we do the oxygen walk, and we did. The reason for this oxygen walk was to determine if my oxygen level was good enough before the discharge and to establish if I was qualified to take oxygen equipment home. At the end of it all, we resolved to order one for rental. My husband ordered oxygen as he was directed. I needed oxygen at home and there was no way I would be allowed to go home without one to call mine. Now that oxygen has been eliminated from the list. The doctor insisted that I must have bowel movement because she wanted to ascertain that there was no blood anywhere.

I bet my sickness had gotten the better part of me. I was done and needed to get going. My nurse's shift ended too but before she left, she introduced the next nurse to me. The nurse came to my room with a suppository and said, "The doctor wants me to insert this in you because she needs you to have a bowel movement."
I responded, "Nurse, please, don't say that. I can go use the restroom if you want me to."

Then, the nurse jokingly said, "Then do it and everyone will be free."
I replied, "Watch me use the restroom."

The nurse kept pushing it because the doctor said she should convince me, but I vehemently turned that down. I had a huge problem with it. I already explained that the reason I have not had a bowel movement was because I had sparingly eaten and not because of constipation. I was speaking out of my experience since this infection. Whatever the situation, we should let the liquid I drank already take care of it. The problem escalated as the nurse came with these four sachets of tablets and asked me to take them all at a time. She went further to explain that the doctor confirmed through my lab results that my sodium was so low. After being stubborn for a short time, I had to take the tablets. I was a little apprehensive here because there were four tablets. I took it but refused the suppository.

I spent a couple of hours with this new nurse before my discharge. It was tough. I was waiting to be discharged, but according to the doctor, it was conditioned on two things: having oxygen at home, which my husband had procured, and having bowel movement. On the bowel movement, I told the nurse to watch me have one without any insert. I was so surprised they went that far after I had told them the reason, I simply told the nurse to be ready while I went to the restroom and had a bowel movement immediately without any constipation. It took only a split second. She came to the restroom, inspected, and confirmed that it was perfect. Now, we've gotten oxygen, I had a bowel movement, and the doctor sent my discharge documents.

I had to wait for my husband. They related everything concerning the discharge proceedings to him on time but no one knew for sure when all things will break even. He was privy to all the happenings. The doctors kept good communication with him throughout. Waiting for my hubby is not my main problem but the bowel movement refused to stop. It kept going on and on. I had to snap at a point when I did not see the nurse. I had to ventilate my feelings to her over the phone. She came and took off the IV on my hands and asked that I let her know when my hubby gets through with the pharmacy. I called her after my hubby informed me that he was done with the pharmacy. She got me ready to go, brought my things out, and got two people that wheeled me downstairs. I joined my hubby in the car, fixed my oxygen on, and we left for home.

This was my third homecoming indeed and my kids were happy but skeptical this time around. I tried to convince them that all is well. This ended up being the last admission because of this deadly virus. My kids were excited that I came back but were curious to know my real state of health. They wanted details. While I was glad to be home, I had no stamina for details. Everywhere was so neat and sanitized exactly how I love it. I felt so comfortable and well taken care of like they did at the hospital. My family gave me excellent care. They took my food orders too. The only snag is that I despised the new addition to our family, oxygen equipment. This thing made so much noise like a small electric generator, and it was constantly on. Like the doctor instructed, I had to be on it for two consecutive days but that was the last I was on that machine till date. It reminds me of bottomless pit because it echoes like a huge empty barrel. We called for it to be picked up after a month of no use. I came back from work one day and it was gone, and I was happy. I am eternally grateful.

I healed every day and so fast too. I continued my breathing exercise and minor exercises. I did not have shortness of breath after my last hospitalization. I got a series of calls from Kaiser and other teams checking on how I was recovering. It felt good that they followed up with me. I gently joined my daily routine with much caution. It was at this recuperating period that the thought of writing came to me, and I embraced it. The whole episode was like a dream, but this is how I beat COVID-19. Isn't it something to be ecstatic about? I am greatly thrilled. I was completely healed by my Maker.

My daughter was so helpful with the new blood sugar spike. This fifteen-year-old became a professional blood sugar manager that I became so petrified with pride. How did she get the hang of this so fast? The diabetes nurse spoke to her and my spouse on my first discharge. She carried out the duties so well. Her bookkeeping was so detailed and easily understandable. I rated her service five stars. As the spring break ended and it was time for her to go back to her school hybrid system, I said to her:

"Mama (I call my daughter all kinds of names), it's time for you to return to school."

She responded, "No mom, I will do zoom so I stay home and take care of you."

I insisted, "You have to go back, Baby, I promised your teachers you will resume after the break."

Reluctantly, she said "If you say so mom. I'm not happy though."

And I answered, "It's okay. I will be here waiting for you to come back."

She told me that I needed her assistance, but I insisted that she should resume hybrid schooling. She did my blood sugar every day before school, after school, and before bedtime. It was a very tough exercise. It was even tougher for us who had never experienced diabetes until this virus infection.

When I got better, I really wanted to relieve my daughter of this task. As much as I detested the sight of blood and injecting myself with insulin, I felt that it is wrong for my daughter to be doing that either. So, I braced myself and decided to take control of things. I started by poking myself and closing my eyes, then went to shooting insulin in my tummy. I can still feel the needle piercing into my tummy. I persevered and learned patiently till I got the hang of it like my daughter did. I took over that task. If anyone had told me that I could do this diabetes thing, I would not even lose a saliva in response. I did it and I am glad that I broke that jinx of not standing blood at all. I still do not like its sight but I'm better now.

I stayed on task, monitored my blood sugar and with the help of my husband, I embarked on a good diet journey. I also embarked on a minor exercise spree since I could not do major ones at the time. I put forth great efforts because I really wanted to reverse my blood sugar status after COVID-19 infection. I also had these weird sensations on my fingertips because of constant pokes. I felt I had no other choice than work on it. The diabetes team called several times and was pushing me to use insulin very often, but I told them to give me time to deal with this naturally. I did not want to rely on insulin except my body does not produce insulin. They gave me two weeks and I accepted the challenge.

I monitored my blood sugar and diet until I had no need of insulin. My level went down drastically till my scale fell to normal range. My joy had no bounds the day I was told to stop using insulin. Yes, I did it and I am so grateful to God. I crossed this milestone so fast. It was not easy, but it was attainable. It was tough. I had to train myself to start liking healthy food as a diabetic. I adapted easily with the diet though, but the squabble was with the blood sugar checks. I now eat healthy even with my blood sugar normal. My new resolution since I recovered from this ferocious virus infection is, "eating healthy."

With this little experience on diabetes, my heart goes to any human that has to daily pass through this unpleasant situation. My thumb is up for you for being steadfast to this course. I went through it, but a smile is needed to live through it. No matter how hard and how uncomfortable this might be, smiles win this race. Smiles make it bearable. I smiled through it and trusted God for complete healing, and He showed up. My heart is with you and my thumbs are up for you.

I was still going through the healing process. I was worried and pitied my weak lungs because I did not understand how the lungs are configured. I thought that my lungs would forever be weak because of this infection until my husband and my friend vividly explained the nature of the lungs to me. Those explanations really helped; they hastened my healing. I was so liberated from my earlier notions and I embraced the facts that lungs can get strengthened and stronger with exercises. On learning that I could help my lungs heal faster, I had to put forth great efforts in achieving that. I sped up my breathing exercise and laid face down every night like I was advised. I became very intentional on how I handle my lungs and I believe it healed fast. I could attest to it, judging from the levels I got when I checked with my oximeter. They were excellent and very normal like I did not go through any of these ordeals. I have no doubt about my healing. The shortness of breath varnished abruptly without trace. I did not have any reason to contact the doctor. We waited for a month and the doctor signed for us to return the oxygen since I had no need for it. God is faithful forever.

CHAPTER 4
MY OUTLOOK ON LIFE: POST-COVID EXPERIENCES. OPINIONS AND APPRECIATION

On God

Having COVID-19 infection and my healing from it, demonstrated how magnificent our God is. Just the simple faith I bestowed on Him, gave my spirit the rest it needed while I was going through this ordeal. The thought that He got me in His hands gave me the courage to face this virus attack without much fret. How God weathered all my storms is still a miracle to me. The miraculous way my problems were solved, made me know that God's hands were strongly in it all. I got absolute and total healing from God. Not even all the side effects from medications were factors in my case. Not even shortness of breath persisted after healing from COVID-19. I had no reason for any physician's assistant after leaving the hospital. God healed me totally and wholly and it is permanent. My soul magnifies the Almighty God.

God is a prayer answering father. The mighty One in battle for that matter. Family and friends prayed and stirred the heavens. They interceded on my behalf and God showed up big time and proved His omnipotence. His praises will never depart from my lips. Wow, I felt so loved by all. I applaud Him for not forsaking the prayers of His people. He answered each and every of the prayers, my healing remains a testimony. He gave me great favors and complete restoration.

He gave me dedicated doctors and nurses that had just a common motive to save me. My heart will forever be grateful to all my family and friends who devoted their one second, nights, days to lift me up to God. You all are amazing, all your supplications to God concerning my life did this healing magic. May you all find favor in the Lord just like I did whenever you are in need. May helps and prayer warriors take up your burdens at the most crucial time in your lives. God's name will be praised forever and ever for great things He has done. I will sing of

His praises forever because He gave me a testimony, and nothing will come between me and this testimony. I lift Him above all the heavens and the earth. Thank you, Lord. I will rather testify to God's glory than have people converge together to celebrate my absence from earth. God broke the covenant of death in my life and set me free. All adoration be His forever. Who am I that you are so mindful of me, my Lord?

On Humanity

Where do I start? As an individual, I have always loved people a lot. I take special steps to show love, it's an intentional habit. I go extra miles to give people breaks. I let humans be human. What I find difficult to condone is disrespectful and grossly irresponsible characters. I have always known that people love me to bits too. They feel comfortable around me. I make conscious efforts not to be judgmental. My experience after COVID-19 has given me reasons to love people even more, without discrimination. I saw angels in human forms at the period of my ailment. I saw and felt love from people, even as a COVID patient.

I believe that God sent me doctors and nurses in the forms of angels to attend to me. The attention I received from Kaiser hospital at Antioch, California was pleasantly overwhelming. These nurses were immensely amazing. They were on my case like they had known me before my admissions. I am not talking about one or two nurses, I meant, all the nurses who attended to me. They were phenomenal. I learned that I should elevate my hospitality game because I was thinking that I was good before my experience. I was clothed with God's grace and bathed in His favor.

I am still in awe. I now have a different view about race though I have been careful on this issue. I will not be fast to conclude things on issues based on race without a fair thought. Race is a personal thing, these nurses in Kaiser, Antioch, that assisted me were from different races. They gave me care like a family. I could still remember how two of the nurses, Melody and her friend, snatched my socks away, moisturized my feet, massaged them, and replaced the socks with a new pair. The other nurses were great too, but this action captivated my interest specifically. Meanwhile, I was hiding my legs so no one could see how dry they looked, not that the nurses were aware. They did not see my feet, but they felt like changing my socks. What they saw might have prompted their subsequent actions. The way they got me new clothes and helped me change, gave me great joy. All these played against my prior knowledge.

I remember my friend telling me how COVID patients are being treated at the hospitals, but I did not experience any of that. I had nurses spending over 40 minutes straight in my room, depending on what activities they had to accomplish. I have not noticed any of them rushing to get away. I remember asking a nurse who spent a really long time in my room what she was doing, and she said that she was pulling out medications. They did their work on the computer sometimes, or other things or were busy attending to me. I've not noticed any of them running off.

They were nice and very compassionate. They asked me a series of personal questions. I could remember nurse Melody asking,
"Do you have kids?" I responded, "Yes. Two teenagers, a boy and a girl. What about you?" I asked. She replied, "No, I don't but I have nieces and nephews"
She asked again, "Are your kids as tall as you are?" I smiled in response, "They are taller than me." So, she said, "You are so tall, share with us." I just laughed and said, "I'm not that tall, I'm just 5.11." They just say things to cheer me up.

The doctors were amazing too. I could still remember how one of the doctors talked about me excitedly and made the nurses giggle.
She exclaimed. "I kept looking at you. Who are you?" I did not understand, and I answered with a confused look, "What doctor, I don't understand?" She responded, "You are a COVID patient, and you look like this, which planet are you from?" Still confused, I asked, "I still can't understand, Doctor." Looking so straight into my face, she explained, "I mean, COVID torment their victims and leave them totally beaten before leaving them, especially

patients who had severe attacks like you. But here you are, about to be discharged, yet looking untouched." She really said much with seriousness. The funny thing is that this Doctor remained the only one there that reminds me that I'm a COVID patient. Even with her gear, she made sure that she gave me safe distance before unleashing on me. All the nurses around laughed loudly. She said that she couldn't believe that I was about to be discharged looking the way I did. I only gazed while others laughed.

I was not finding anything funny as at the time, but I was happy to know that I was doing great. I am sure I was grateful with the compliments but wasn't in the right frame of mind to acknowledge that. Neither was I ready to flow with any joke. Meanwhile, this was the same doctor that talked about my diabetes. She was the doctor that made sure I followed every instruction; she is a great doctor. She was on top of my case, and I appreciate her genuine concern. I can still hear her voice telling me to stop being in denial.

I was cautious of the contagious nature of my ailment and tried hard not to infect the people attending to me. I made sure my mask was constantly on. They felt for me and wanted my mask off. I have had a nurse practically cut off my mask with scissors and voiced her reason. She believed I shouldn't bother myself wearing a mask since they come in covers and in their gear. Some told me to just take my mask off and some said I should take my mask off once they leave the room. To be precise, they were all nice and categorically, meant that I should not bother myself. I saw people who did not want me to deal with anything at all except my ailment and recovery. I am really bereft of words, but I am grateful. These nurses were incredibly exceptional. I still owe it all to God's favor.

The most touching thing these nurses did was the cards they signed and sent to my home. The pictures are in the gallery pages of this book. The first one hit me so hard. I was unquestionably exhilarated. That kind gesture really quickened my recovery. I was busy thinking on how to show some appreciation for their kindness and they were there busy concocting more benevolence. The first card came three days after my discharge and the second card came about two weeks afterwards, signed by another set of nurses. I cannot emphasize this enough; these nurses were amazingly incredible. I felt loved, a COVID patient like me. God is so mindful of me, just for nothing. I sincerely do appreciate all the doctors and nurses that took care of me at Kaiser Antioch, CA and at Walnut Creek, Emergency, all in California. You all are heaven sent to fulfil this miracle. Thank you for being part of my recovery story.

I had to vividly narrate all these kind gestures because these were solicitudes that were melted on me by people who were neither obligated nor wanted anything in return from me. They neither know me, nor do they know if I will even survive. I am full of appreciation. I cannot thank them enough. I eventually sent my hubby a card and some chocolate to register my gratitude. They are diamonds, I can't treasure them less. I learned a great lesson on kindness.

On Life Experience

I learnt a lot from being sick with this fatal virus. I chose to share my experience and leave words of advice as a result. I believe this might impact someone positively. The energy we send out sometimes attracts the response we receive.

I am from a culture where people hide a lot of their medical experiences. The norm would have been for me to claim that I had just a minor sickness. My culture does not promote sicknesses. We hide them, but the complication is that those illnesses do not reciprocate our secrecy, they do not hide us. I am not promoting COVID here because there is nothing good about it. I loathed it to the level that I hate to spell it out with uppercase letters. Some people will still get highly irritated with this book because to them, I am letting the world into my privacy. Well, I am packaged differently. I felt that I must share this for so many reasons. This is my testimony to God's awesomeness in my life. These are my past experiences which can coerce someone out there to make a positive decision on this virus or even on other life experiences. I go as far as narrating my ordeal to anyone that hints me about not taking the vaccine. I let them make a choice after hearing my story. I have had people I encountered get their vaccines after my story. I spoke to someone who had one shot and ran away and she went for the second shot.

I had to share my story so that people who are still in denial concerning this virus should re-examine their resolutions. I shared because I want to proclaim the goodness of the Lord in the land of the living. I believe He put this testimony in my mouth. I share because I marveled at how kindly people who were not even of the same race with me attended to me at the hospital. I met people who are the same race with me too, they treated me so nicely too and I do appreciate them. I did not say much about them because I expected that. I shared because I am overwhelmed with the love I experienced from all sectors, my family, friends, the hospital, etc. One thing I must say here is that there are excellent nurses and good people out there regardless of race. A

Now, I pay more attention to details on things I usually would take for granted. I am currently more sensitive to my environment and show empathy at what people are going through. Going as a diabetic for six weeks, poking and giving myself insulin taught me that we should never conclude on things. Good health should never be underestimated. I would not let them mention that while I was at the hospital. I refused to learn about diabetes because I understood I was not diabetic. But I had to learn and relieved my little girl of that duty. I wanted her to be the child she is. That was my cross, not hers. So, she mentored me on how to check my blood sugar and how to inject myself with insulin and I became a pro at it. I am so thankful to God that my blood sugar came back to normal. Gratitude will always be my attitude. Now, I can discuss with anyone from my personal experience. I can advise someone that needs such help.

These six weeks of being diabetic was supposed to be very miserable, but I combined it with lots of smiles and thanksgiving. In those six weeks, my average daily pokes to check my blood sugar was about five times. I did not mind, all I wanted was to heal from COVID-19 infection and its cohorts. Now, I know what people go through when they are diabetic, and my heart is with anyone who goes through these blood checks every day. Please, make sure that you add some smiles to it. Smiles make tough situations subtle. I believe it is the only way to make meaning out of it. Import some measure of discipline too and you must surely triumph. Did I say that I started enjoying sugarless dishes? Yes, I fell in love with vegetable smoothies. I complained any time my husband added two strawberries to make it tasty. To me, there's always beauty in every ugliness, just look so closely and think so deep.

On Situations

Going through this agonizing experience taught me that tough situations strengthen us no matter how weak we may be before the depressing situation. I was in that lonely COVID room without any fear. I could not believe that I could overcome fear in isolation. I became a spirit myself when I was facing this devil. Usually, I am scared of being in a hotel room all alone. I have never slept in a whole house alone. I never let that happen in my life. On a normal time, all I will be thinking will be all about the ghost getting at me. I still could not believe that I slept face down without thinking that the ghosts would come from behind and get me. A different situation for real. There I was, I remembered that so many people have died because of this virus and possibly, in that same room. Invariably, people must have passed on in one of the three rooms, if not all the rooms I stayed in while I was on admission. I could not believe that I accepted the advice to sleep face down while I was alone. I cannot still fathom how that could have been possible. The days of constant use of 'Never' is officially over for me, thanks to COVID-19. Experience is an excellent teacher.

It was delectable indeed for me to know from this situation that I can count on my friends. I am still so thrilled on how friends rallied around me at the least notification and pulled down the rains. My friends have been there for me on other instances but their expeditious response on hearing about my COVID infection is highly commendable. All the prayers on behalf of my family and I were unbelievable. Friends and family petitioned to God and He heeded. I felt the love. To hear some of my friends cry over the phone on learning that I was diagnosed with coronavirus was heart melting for me. These experiences are indeed life changing. I was touched deeply and learned that certain situations need immediate reactions. If I did not have a situation like this, perhaps, I would never have come to this realization. I appreciate all my family and friends for all their love, prayers, and other contributions. We all were in this together, we beat COVID-19 hands down.

It is always good for situations to arise, so we widen our knowledge. I never knew that my hubby could handle stress the way he did at that trying period. He absolutely handled it squarely even though he was also diagnosed with COVID. He was the cook, the doctor, the janitor, the cab driver, etc. I relaxed and savored his care and attention. He sure made some lemonade with the lime life threw at him. I am still in awe of how he perfectly managed the situation and perfectly did his office job from home as well. His patience at that time was highly commendable. He got two thumbs up.

That perilous time proved my kids' level of discipline too. They were calm and obedient to instructions. They behaved like adults, very understanding. They were positive to COVID-19 too with mild symptoms. My daughter took my breath away on the maturity she exhibited throughout the period. She was 15 years old, but she handled my steroid induced diabetes professionally. I would have attested that she could not face tense situations like mine, but she aced this one. I guess we will never know what life's situation might throw on our laps.

The reversal of events regarding my attitude to medicine after my rendezvous with COVID-19 was 180 degrees. I hated medicine like plague. Usually, when I am sick, I always find excuses to avoid medicines. I became a changed human in this COVID-19 situation. I did not play with these medications. I did not miss even a single dose. Nothing came between myself and my medications till they were all gone. I guess I was very eager to recover. So, all those while I frowned at medication must have been that I have not experienced a severe situation that leaves me with no other choice.

Injecting myself was the apex of my encounter with COVID-19. Prior to this time, I usually shut my eyes any time I got mine or my kids' blood drawn in the lab. I despised the sight of blood. COVID infection got me in closest proximity with blood through checking my blood sugar and shooting insulin when necessary. It took me some time to get the hang of it, but I eventually did. I feel like I can do anything now. No one could save me from that ordeal, so I threw my worst fear to the wind and got down to the blood sugar business. Thank God, I got my normal blood sugar level back. We should not say 'Never' because our situation might invite our worst nightmare.

I came to the realization that good people exist no matter their race, belief or culture. All the medical crew that attended to me were superb irrespective of their different religions, I still felt their kindness. I also came to the knowledge that Kaiser respects every religion. They asked me my faith the first day I was admitted, I said Christianity, and someone called me later on in my admission and prayed with me. I was indeed happy that such transpired. Being sick with COVID-19 unquestionably widened my experience on life generally. My COVID situation and experience taught me to see individuals as individuals first, before questioning their religion. Religion do not make a person but it can help mold a person's morals.

CHAPTER 5
CONCLUSION

VACCINATE!

All these said, it's very pertinent to add that we should all hasten our efforts in getting the vaccines. I will not wish this virus on anyone. I bet, it's better to get the shot and have mild adverse effects and be healed than going through the whole nine yards of this disease like I did. To me, there is no comparison between the vaccine and the infection. The earlier we get our shots, the better for us and the entire world. Do not wait, go get vaccinated. Make hay while the sun still shines. Even as a person who has suffered from this virus, I have antibodies according to science, but I verified from my doctor about getting a vaccine and immediately he gave his approval, I went for it. I got my shots, the two doses. I know with experience that whatever side effects that come because of the vaccines can never be compared to COVID-19 infection. Personally, I did not experience any side effects except for mild pains in my arms for a couple of days.

Get your vaccines, you will be so gratified you did, I promise you. If not, tell me how you will cope when people discard their masks?

Did I mention that my business suffered greatly because of the deadly infection? It was simply a disaster. All these while I was going in and out of the hospitals, The Beauty Paradise, women clothing store was going through its own sick time too. I had no competent hands to keep my shop open. I certainly do not have trust issues. Like I said earlier, I give humans a break. I have a lady, Shelley, who works on contract basis with me but she was unavailable at that time. When I called her to take care of my business in my absence, she said that she was not in the vicinity and I believed her. She loves working with me.

Unfortunately, she is the only person who understands what I do at a close range. So, I locked up my store and followed my destiny. That took over a month of closure, but I didn't bother about it. As usual, I surrendered it in the hands of God. I reopened as soon as I felt a bit better, although staying home a little longer would have been preferable. I had to do the needful. This is another thing I could have sworn will never happen to my business. Being around the Bay Area and my business closed for over a month without any lockdown. Now, I know that anything is possible in life.

For those who still doubt this virus, please, heed to someone who already has first-hand experience. I have been there, and it is obviously not the best route. Get your vaccines now you still have the grace, do not experiment with yourself. A wise person learns from someone else's experience. I have shared my experiences, so you do not experience it before taking precautions. I went, I saw, and I conquered but it is not always the case. All glory to God, The King.

APPENDIX:
PICTORIAL GALLERY

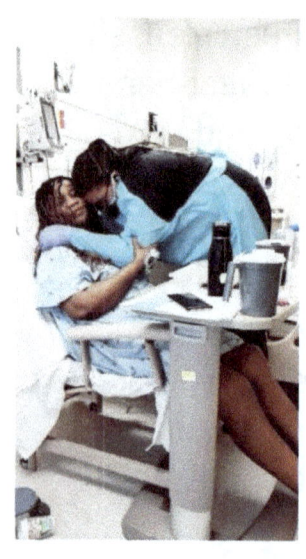

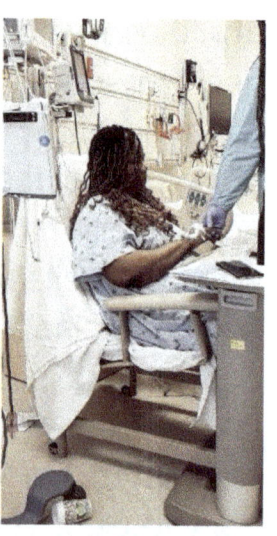

(All four pictures were taken on the day of my discharge. My husband and daughter were invited to take training on how to care for a diabetic and also to take me home after several days of being in the emergency.)

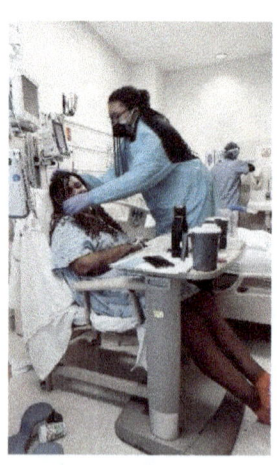

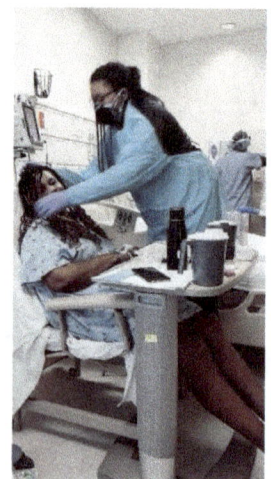

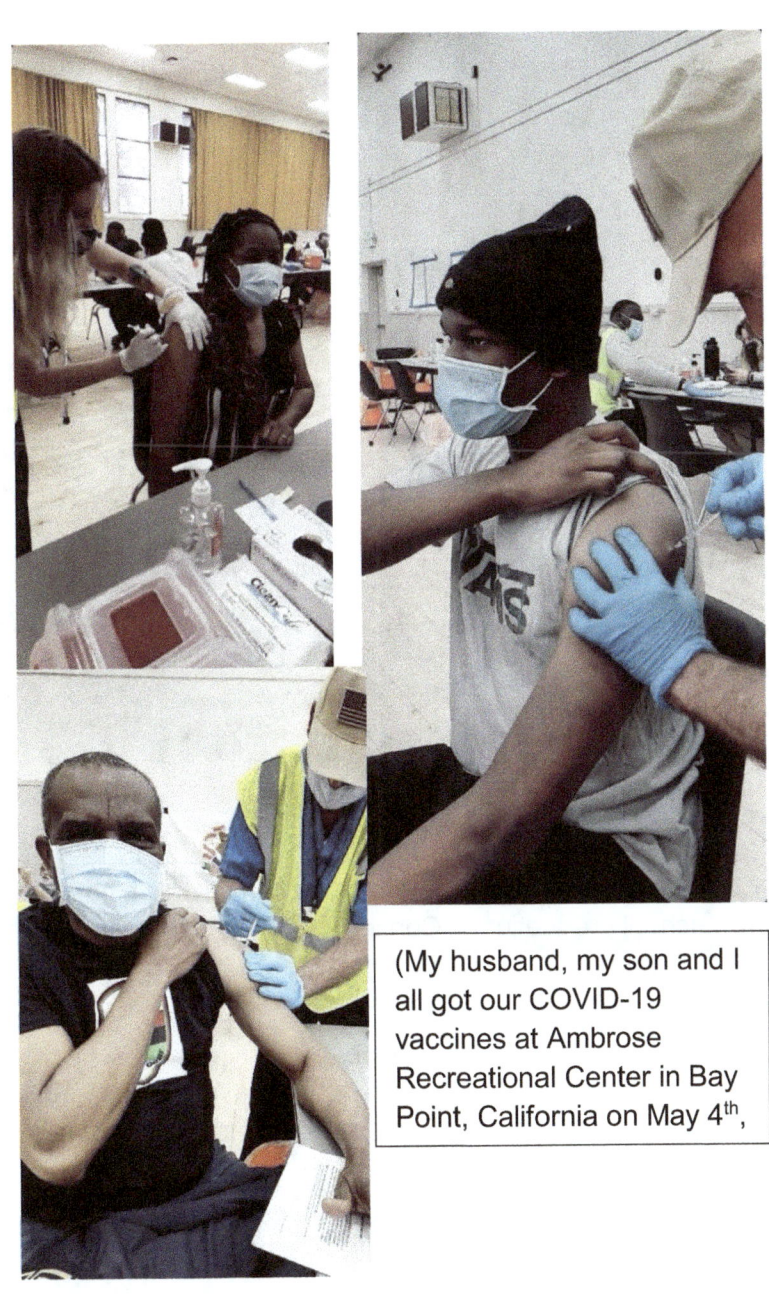

(My husband, my son and I all got our COVID-19 vaccines at Ambrose Recreational Center in Bay Point, California on May 4th,

My family and I on the 4th of June at my son's high school graduation at Concord Pavilion, California.

First picture was taken on March 6th, 2021, just before my COVID diagnosis and admissions.

Second picture was taken on April 23, 2021, that was my first picture after my COVID admission. Still with the same

(My daughter's 16th birthday pictures in April. About two weeks after my last hospital discharge.)

This card was signed and sent to my home three days after my discharge by nurses in the COVID team of Kaiser Permamente, Antioch, CA. This card meant the wordld to me

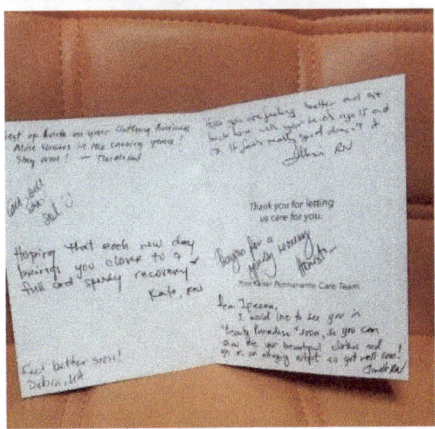

The second card also came from the Kaiser nurses two weeks after my discharge. This one blew my mind away. I felt loved. I am eternally grateful

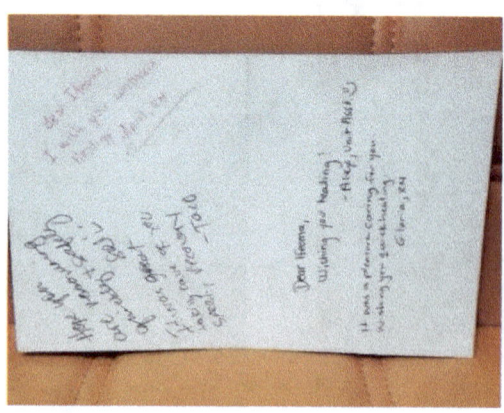

www.ingramcontent.com/pod-product-compliance
Lightning Source LLC
Chambersburg PA
CBHW070808220526
45466CB00002B/594